I0817123

Natural Resources

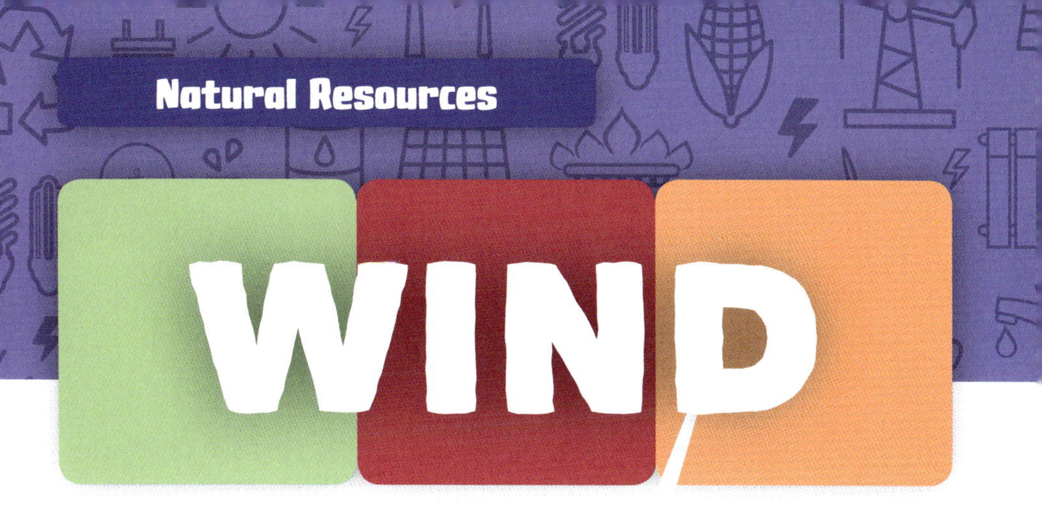

WIND

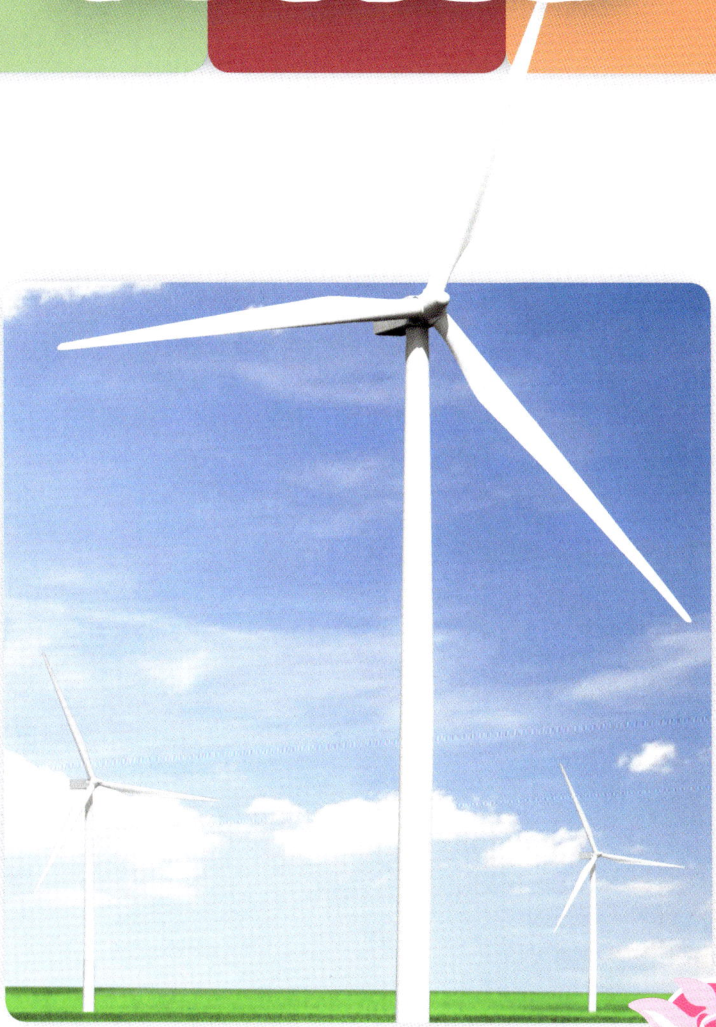

Gail Terp

DiscoverRoo
An Imprint of Pop!
popbooksonline.com

abdobooks.com

Published by Pop!, a division of ABDO, PO Box 398166, Minneapolis, Minnesota 55439.

Printed in the United States of America, North Mankato, Minnesota.

102019
012020

THIS BOOK CONTAINS RECYCLED MATERIALS

Cover Photo: Shutterstock Images

Interior Photos: Shutterstock Images, 1, 5, 15, 16–17, 18–19, 21, 22, 23, 27, 30, 31; iStockphoto, 6, 7, 8–9, 10, 11, 13, 14, 16, 24, 29; Ashley Cooper/Alamy, 28

Editor: Sophie Geister-Jones
Series Designer: Jake Slavik

Library of Congress Control Number: 2019942629

Publisher's Cataloging-in-Publication Data

Names: Terp, Gail, author.

Title: Wind / by Gail Terp

Description: Minneapolis, Minnesota : Pop!, 2020 | Series: Natural resources | Includes online resources and index.

Identifiers: ISBN 9781532165900 (lib. bdg.) | ISBN 9781532167225 (ebook)

Subjects: LCSH: Wind--Juvenile literature. | Wind energy--Juvenile literature. | Natural resources--Juvenile literature. | Environment--Juvenile literature. | Ecology--Juvenile literature.

Classification: DDC 333.92--dc23

WELCOME TO DiscoverRoo!

Pop open this book and you'll find QR codes loaded with information, so you can learn even more!

Scan this code* and others like it while you read, or visit the website below to make this book pop!

popbooksonline.com/wind

*Scanning QR codes requires a web-enabled smart device with a QR code reader app and a camera.

TABLE OF CONTENTS

CHAPTER 1
WIND AT WORK

It is a windy spring day. A girl watches her yellow kite soar in the sky. Nearby, a group of kids set sailboats in a pond. Wind fills the sails and pushes the boats

WATCH A VIDEO HERE!

Sailboats catch the wind in their sails and use it to power them across the water.

across to the other side. This is just some of the work wind can do.

Wind happens when air moves. As the sun shines down on Earth, it warms the air. Some air is warmed more. Warmer air has lower **air pressure**.

DID YOU KNOW?

When low and high pressure areas are close together, wind is faster. When they are far apart, wind is slower.

Kites require wind to stay in the air.

As a result, the warmer air rises. Cooler air has higher air pressure. This air falls. It moves to where the warm air had been. This moving air is wind.

Wind moves weather from place to place. Wind blowing from the freezing **arctic** brings cold weather. Wind blowing from a **tropical** place can

People can watch the wind move storm clouds across the sky.

bring rain. In addition, wind blowing over water gathers **moisture**. This moisture forms clouds. When clouds have enough moisture, rain falls.

HOW WIND IS MADE

COOL AIR WITH HIGH PRESSURE DROPS.

THE MOVING AIR BECOMES WIND.

WARM AIR WITH LOW PRESSURE RISES.

CHAPTER 2
POWERFUL WIND

People use wind power to do work. In the past, people built windmills to grind grain. Wind turned the windmills' blades. The blades turned a rod. This rod turned a stone that ground grain into flour.

LEARN MORE HERE!

The Greek island of Mykonos is home to windmills from the 1500s.

Windmills also helped do other tasks.

Some pumped water up from the ground.

In the 1880s, people began building **wind turbines**. These machines use wind to create electricity. First, wind spins turbine blades. Behind the blades, a long rod connects to a **generator**. The spinning blades turn the rod. The turning rod causes the generator to produce electricity.

WHAT MAKES A WIND TURBINE

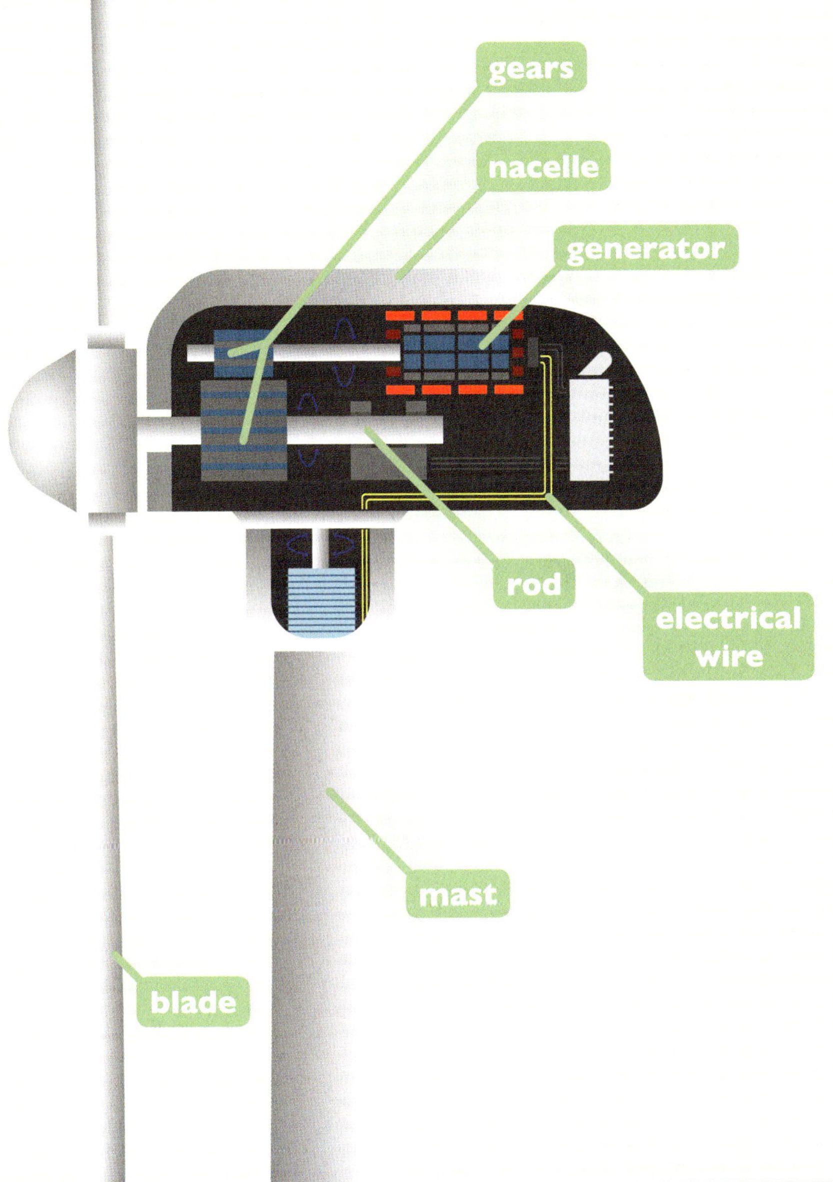

Today, wind turbines come in different sizes. Smaller turbines can power homes, farms, and schools. Larger turbines are grouped together on wind farms. Wind farms create huge amounts of electricity. They provide power to large areas, such as cities.

Wind turbines are made up of approximately 8,000 parts.

The nacelle is the part of a turbine that holds the generator. Some nacelles can be 50 feet (15 m) long.

Wind power has many **benefits**. Unlike some of Earth's resources, wind will never run out. It does not get used up like oil and coal. These fuels must be

Industrial wind turbines have blades that are approximately 116 feet (35 m) long.

burned to produce power. This process creates **pollution**. But wind turbines do not pollute.

CHAPTER 3
A FEW PROBLEMS

Wind energy does have some downsides. Not all places have the right type of land for **wind turbines**. Most wind farms are built on flat, open land. Tall trees and forests can slow down wind's flow.

COMPLETE AN ACTIVITY HERE!

Many wind turbines are built on prairies and farmlands.

Mountains can also block wind. Or they can have gusts of very fast wind.

Even when wind has nothing blocking it, it's not always dependable. It may stop blowing, or it may be too weak. Without strong and steady winds, turbines can't produce enough power.

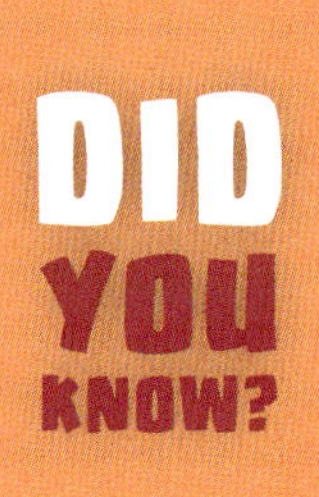

The fastest wind in the United States was recorded at Mount Washington in 1934. It reached 231 miles per hour (372 km/h).

High in the mountains, gusts of wind can suddenly blow very quickly.

The speed of a turbine's spinning blades can affect the air pressure around them. This can harm animals such as bats.

Wind turbines can cause problems for wildlife. The blades spin so quickly that birds and bats don't always see them. If the animals fly too close, they can be hurt or killed.

CHAPTER 4

THE FUTURE OF WIND

Scientists are working to solve wind power's problems. Some places use offshore wind farms. These **wind turbines** are built over water. They let people use wind power along coasts.

LEARN MORE HERE!

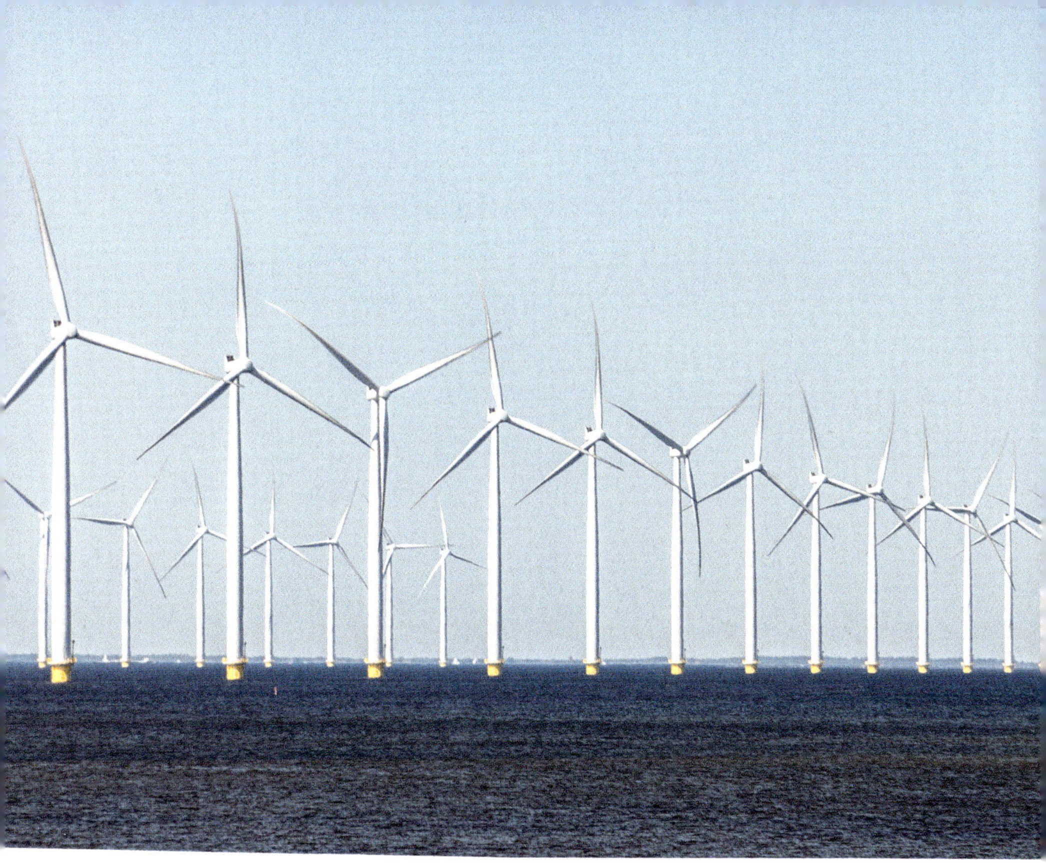

Offshore wind farms catch breezes blowing over lakes and oceans.

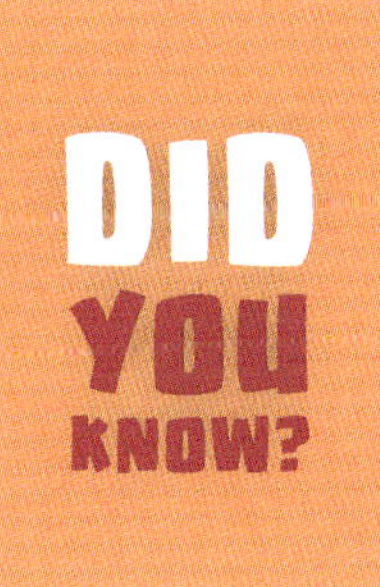

Wind power is used in countries around the world. The top three countries are China, the United States, and Germany.

Scientists also work to make turbines cost less and create more electricity. Some new turbines use batteries. The batteries store electricity.

Batteries are stored at the base of the wind turbine.

Jet streams are currents of fast winds that blow very high above the ground.

KITE POWER

Wind that is high above the ground blows stronger than wind closer to the ground. Scientists want to use this strong wind to produce power. Some hope to use kites. The kites will fly high in the sky. Their lines will connect to turbines on the ground. As the kites fly, they will power **generators** to create electricity.

New turbines also spin more slowly. Slower turbines are quieter. They kill fewer birds and bats.

MAKING CONNECTIONS

TEXT-TO-SELF

Would you want to live near a wind turbine? Why or why not?

TEXT-TO-TEXT

Wind is a renewable resource. It does not run out. What other renewable resources have you read books about?

TEXT-TO-WORLD

People use wind power to produce electricity. What are some ways you use electricity at home or at school?

GLOSSARY

air pressure – the weight of air as it pushes down on Earth.

arctic – the area around the North Pole.

benefit – a good or helpful result.

generator – a machine that converts motion into electricity.

moisture – the water in the air or on something.

pollution – harmful substances that collect in the air, water, or soil.

tropical – warm and wet.

wind turbine – a tower with large blades that is used to produce electricity.

INDEX

ONLINE RESOURCES

popbooksonline.com

Scan this code* and others like it while you read, or visit the website below to make this book pop!

popbooksonline.com/wind

*Scanning QR codes requires a web-enabled smart device with a QR code reader app and a camera.